Cryptocurrency Trading

A Complete Beginners Guide to Cryptocurrency Investing with Bitcoin, Litecoin, Ethereum, Altcoin, Ripple, Dogecoin, Dash, and Others

Crypto Tech Academy

advice. The content of this book has been derived from various sources. Please consult a licensed professional before attempting any techniques outlined in this book.

By reading this document, the reader agrees that under no circumstances are is the author responsible for any losses, direct or indirect, which are incurred as a result of the use of information contained within this document, including, but not limited to, —errors, omissions, or inaccuracies.

Contents

Dibbly Publishing

Dibbly Publishing publishes books that inspire, motivate, and teach readers. Through lessons and knowledge.

Our Book Catalog

Visit https://dibblypublishing.com for our full catalog, new releases, and promotions.

Follow Us on Social Media

Facebook - @dibblypublishing

Twitter - @DibblyPublish

Download Your Bonus:

Bitcoin Profit Secrets

Discover the methods and techniques used by the most successful Bitcoin investors so you too can profit and succeed!

https://dibblypublishing.com/bitcoin-profit-secrets

Preface

It is as straightforward and simple as it sounds. Cryptocurrency trading is just the act of buying one cryptocurrency in exchange for another, or using fiat money (like the US dollar) to purchase cryptocurrency. Just as you would go to the exchange or to a bank that provides the service to convert your currency when you travel abroad, you could visit a cryptocurrency trader to exchange your dollar for any one of the cryptocurrencies that are in the market.

Essentially you are looking at an exchange-driven transaction between the various cryptocurrencies and the dollar, as well as between one crypto and another. Just as in all financial transactions, there is a trading opportunity that exists when a natural demand between one currency and another exists. The underlying currencies, let's say the bitcoin and the dollar, have a natural rate of exchange between the two. This is the activity of the buyers and the sellers around the world who need to conduct their business in Bitcoin (BTC) and still have the US dollar in their hands (USD). When they enter the market and make the purchase, they create demand.

When a person who receives BTC as payment for goods or a service goes to the exchange, they are desirous to sell the BTC and get USD for it. When they go to the market and sell BTC, they are creating a supply for it. As with all demand and supply, the balance between how much is demanded and how much is supplied is eventually reflected in the price.

In July 2010, the market-determined price of BTC was 8 US cents for every 1 BTC. In July of 2011, a year later, 1 BTC was trading at $34. That's $33.92 appreciation in a year, or 42,000% change. In July of 2016, BTC traded at $600. By the end of 2017, the price of BTC reached above $17,000. What was once just valued at 8 cents was worth nearly 18,000 dollars. At the time of writing, it stands at about $11,000.

When something can be expressed in terms of value, has numerous willing buyers and willing sellers, and has a use beyond the financial markets, what you essentially have is a tool that can be traded, arbitraged and profited from, merely from its financial profile.

Take the US dollar for instance. It satisfies all three of those conditions in the last paragraph. It can be expressed in terms of value against any number of things. Many of the things that you buy, not just at the local grocery store, but also on foreign websites, list their prices in USD. You can use the USD that you earn from your job, received as a gift, or got in return from selling something, and use that USD to buy whatever you want. And anything you want to buy is listed in terms of the price that is quoted in USD. That creates a strong underlying market.

Most sovereign currencies are only accepted within their borders. If you go to South Korea, the won is what they accept there as currency, but you couldn't use that won if you hopped over and visited China. You would have to go to an exchange with your won and purchase the yuan. Because there are numerous people and entities constantly buying yuans and selling wons, a price can be derived, and that is called the exchange rate. It is a willing-buyer and willing-seller mechanism.

This makes currencies tradable between themselves. Unlike stocks and shares in the equity market, the diversity of the Forex market creates opportunities for trade with more frequency and certainty. We will discuss that later.

You've seen this picture before. That's how the trillion-dollar foreign exchange (Forex, or FX) market works. It is the exchange of one currency for another. It is the same with cryptocurrency.

With the emergence of BTC as a currency, it advanced to becoming an asset once there was a large demand for it and there was a correspondingly large supply of it. That meant, to economists and market watchers, that not only was there a large base of buyers and sellers, but there was also a market that was based on speculation.

There are millions of traders who bet (not used as a gambling reference) on the direction of the USD on a global basis. In total, more than five trillion dollars' worth of transactions happen in one day, every day. From this amount, less than 1% is transactional. That means that there is an underlying commercial transaction where one party entered the Foreign Exchange (FX) market in order to buy the currency they needed so that they could pay their vendor. Out of the other 99%, it can be easily traced that 80% is purely speculative. That means traders enter and exit the market with the sole intention of profiting from the purchase and sale of the currency and that there is no underlying transactional reason for the entry to the FX market.

Introduction

As the transactional nature of trading participants changes over time, so too does the way in which the global markets pay for these transactions. In the last fifty years, the USD has been the currency that has driven and been integral to the global market for trade and aid. It has been the world's reserve currency and holder of consistent value. That is why you see international transactions always quoted in dollars. But that wasn't always the case. There is a long history behind that, which is not the subject or scope of this book, but the point to take form this is that the dollar ascended in its prominence over time and because of global politics. It was not always the world's transaction currency.

To get a better understanding of the role and development of currency, you have to understand the transactions that were conducted in the past were done in person or by messenger and written on a piece of paper. The value of the purchase was paid in gold, or bartered and then eventually was traded in dollars.

But times have changed. We are no longer as reliant as we were on bulk transactions, as the international trade market is becoming more liberalized. Further, individuals are able to make purchases on their own across the border and have it delivered in days, wherever they may live.

If you look at companies like Amazon in the US and AliExpress in China, you see that almost anything can be bought and sold and delivered to any point on the global map. That changes the entire financial transaction industry. In the past, you would only be able to go to your retailer at the local town square to make your purchase. He stocked his shelves by buying it from a local stockist, who got his supply from a regional stockist, who got his supply either from the manufacturer or from the importer. The importer would have aggregated the demand and made a bulk purchase from the manufacturer overseas. When that happened, the importer would go to the bank and open a trade instrument so that he could pay for the purchase from his seller. That's the point where the bank would enter the picture and go to the financial markets to buy the currency they needed on behalf of the client.

Now, remember we said that in the world today, only about 1% of the transactions in the FX world are transactional. The problem with this is that it creates wild swings in the price of the currency. There would be few and infrequent transactions and that would cause the prices of the currency to swing wildly. You saw that with the nascent BTC as well, back in 2011. There are two terms you should familiarize yourself with here that will be used throughout the book. The first is "liquidity" and the other is "volatility."

Both the words are not as arcane as they may appear and the concept rests close to the word. Volatility shows how much change can happen as a function of time. When something is said to be volatile, it makes wild gyrations and large price movements with just the smallest transaction.

On the other hand, liquidity means that the asset is able to be exchanged at a moment's notice. Take for instance if the market were illiquid. That means if I were to walk up to the market and look to buy USD, but there were no sellers, I would have to wait for a seller to emerge. In that scenario, if there are no sellers at all but I have an urgency, I would have no choice but to offer someone a higher price. Then I'd have to keep upping my offer until someone is willing to sell me, say, the Canadian dollar (CAD) in exchange for my USD (since I want to pay my Canadian seller). The less liquid the market, the more the price fluctuations (volatility) because I would have to keep raising my offer so that someone finds it interesting enough to make the transaction.

This is a real problem when it comes to international finance. First, there is an unnecessary delay, and second, there is no reliability on prices of currency. That is the main purpose that speculators like you and me play in the entire market system. It takes more than 80% of the total daily volume of transactions to make sure that at any point in the day there is always a willing buyer and a willing seller to come in and make a transaction happen.

These speculators take on the risk of price movements and are paid handsomely for their risk. More importantly, they do an immeasurable and valuable service to the global finance market, by providing a smooth market with minimal volatility and liquidity. When I walk into any market today and try to by CAD for USD, the transaction happens instantaneously because there is always a buyer on the other side of the transaction.

There is one more layer that you should know about. It is called the "market maker." The market maker is the person who will buy from you or sell to you anytime you want, without having to make sure that there is a corresponding buyer or seller on the other side of that transaction. That's because he is making a certain profit from the market-making activity. He takes on the risk of price change in return for the premium that he charges (built into the exchange rate) that you pay for. On the other hand, you get the value of an instantaneous transaction.

Market makers are the important part of the FX world because they close the gap between buyer and seller and hold the currency for the short time instead of you having to wait for the buyer to emerge—even in a highly liquid market. They smooth the bumps and dips and are paid handsomely for their involvement.

Why does this matter to cryptocurrency trading? Because trade, and the mechanism that surrounds it—regardless of whether it is cryptocurrencies or fiats—is still the same. A transaction is a transaction, regardless of whether it is made for hogs or for paper money.

As the world moves away from paper transactions we find that electronic transactions are taking over and becoming more efficient, giving us more speed and versatility. The rate that international transactions now occur is staggering and could not have been fathomed even in the last generation.

The ability for paper transactions to keep up is slowly eroding and the need for a new way to transact is emerging. That is the underlying reason that BTC and other cryptocurrencies are gaining popularity. It's not a fad; it's the response to the new paradigm of business transactions.

There is something that the current structure of currency cannot even do that cryptocurrencies can. Cryptocurrencies have the ability to make any asset into a transactional instrument so they can subsequently be transacted on. We will come to that later in the series, but for now, you have to understand that cryptocurrencies are not just the electronic version of money. It is not just the move from paper to bits and bytes; it's about the nature of the transaction and the speed and security of the transaction within the new framework of trade.

Take for instance the fact that it takes three to five days at the very least to make a payment via check or eCheck. It takes no less than three days to transfer cash, and if you are doing this internationally. it can take three banking days. If there is a weekend involved, that can mean in the region of 5 to 6 days from the date of initiation to the day of receipt. A Bitcoin transaction for payment happens in about ten minutes.

Fifty years ago, without the use of computers, it was acceptable for payment and debt advice to take a month or more while the paper it was written on made its way across the ocean and on to its destination, It was expected that international payments would take weeks to arrive. But in today's world that is unacceptable.

Fraud is another issue. In the case of financial transactions, there is so much opportunity for fraud. There are bounced checks, counterfeit notes, and instances of counterfeit wire transactions. In today's fast-transaction environment, it is important that transaction speed and the security of the transaction is never under the shadow of doubt. Because of the lack of security, transaction participants have to wait while funds are confirmed, and that takes extra time. In BTC transactions, the confirmation happens in minutes. This is an important reason to believe that fiat money will slowly make way for a cryptocurrency system.

These are just some of the things that you should understand before we undertake the task of looking at the benefits and features of cryptocurrency trading.

Why We Need Currency

Cryptocurrency is the latest iteration in civilization's need to monetize transactions. We started out our commercial evolution using the concept of bartering. We moved on to coins that had value of their own—like gold. Then we moved to paper currencies that derived value because of the government's backing.

There are many reasons we need currency, in whatever form that may be. Here are just a few:

1. Storing of value for future use
2. Ability to purchase what is needed when it is needed
3. Ability to place value on other assets

Cryptocurrency provides all this and it provides it a lower cost, greater transparency, and less friction to trade.

Chapter 1

What is Cryptocurrency

Cryptocurrencies, or cryptos, are decentralized peer-to-peer digital forms of monetary exchange. In that sense, it serves as cash where you can pay it to whomever accepts it in exchange for their goods or service. You can also exchange it for another cryptocurrency, or another currency called fiats.

Cryptocurrency is a name derived from two terms—one mathematical, and the other financial. Cryptology is the reference made to codes and ciphers that are virtually impenetrable. It is the stuff of spy films and intrigue. But really, it is about math. There are certain concepts in math that make cryptography a complex science. In essence, it is about coding something for transmission that cannot be deciphered if and when intercepted, nor can it be altered or mimicked. It is almost like a thumbprint of the element or message or underlying content that it is carrying.

Cryptography exists in many forms. If you are using a computer that communicates online in a secure way, the message is encrypted so that only the person transmitting and the person receiving know the necessary code to decipher the transmitted content. The encryption uses cryptography to make sure of that.

The second part of the term is currency. Currency has just come to mean the concept of exchange that is fungible. That means you can exchange it for anything. You can buy a car with it one day, and tomorrow make a donation. The use across almost every area of human living makes currency the most efficient way to conduct transactions. Although people still transact in the barter system where one set of goods is exchanged for another, they inevitably come back to using the value of that transaction in currency. So currency, aside from the convenience of being printed on paper that can be carried or stored as value in a debit or credit card, has the ability to be an assignment of value for anything from cars to bridges, and services to ice cream.

When you put them together, what you get is a system that uses cryptography to encapsulate a transaction and give it a value. By encapsulating it and giving it a value, it can be widely distributed for consumption and it makes the effort to sell it or buy it much easier. But in the execution of this fact, the old model of currency has been found lacking in that the value of that currency could be manipulated by governments. What could buy one ream of paper today, may only be able to buy half a ream tomorrow because the government that is issuing the currency has decided to print more money and thereby dilute the value. One of the things we needed was a way that no single authority could manipulate the currencies of the world and thereby keep things stable—and totally based on market forces of supply and demand.

In today's currency market there are arbitrary pegs and moves to keep a certain currency at artificial levels. There are monetary and fiscal policies of one country that could affect someone two continents away. Decentralized currencies that are used to transact business all over the world do not have that drawback. Something like a bitcoin cannot be printed, nor can the circulation be increased just because some government needs to do so.

That is another feature of the cryptocurrency—based on the guidelines and practices set by the founder and creators of Bitcoin (we still are not entirely sure who that is)—it was designed so that it cannot be increased beyond a certain number in circulation.

Think about that for a minute. You have a currency that is never going to increase in the number of coins in circulation. The only thing that can happen to it is that it is broken up into small fractions of itself. So if 1 BTC today is at $10,000, then 0.1 BTC would be $1000. This has two effects. The first is that no one can manipulate the BTC and change the value adversely by pumping in more BTC. The second is that the demand for BTC will eventually exceed the supply, and that would mean that BTC itself would be valued consecutively higher in time and be fractioned lower and lower.

So for instance, we could end up breaking 1 BTC, into

a thousand fractions and that would mean that each fraction would consequently be valued at $10. This sounds a lot like a stock split. And that is exactly what it is when you think about it. The underlying asset can change in price to reflect the new reality at that point. At the moment, the smallest fraction, or unit, of Bitcoin ever transacted is for 100,000,000th of a BTC, or 0.00000001 BTC. In equivalent dollar value at an exchange rate of $10,000, that works out to be .01 cents.

But the benefit you may realize is then lost because fractionating the currency is the same as printing more money. It actually is not. Printing more money is a decision made by a sovereign. The price of BTC is totally market driven. If the market gets large enough, it is not possible to sway it unless there is a concerted, global effort to sway the market.

All these factors make cryptocurrencies—and the likes of Bitcoin—perfect assets for speculative trading. The speculative trading market, whether it is for FX or cryptocurrencies, is a mutually beneficial situation, for the main reason we highlighted earlier. Speculative players give the market the liquidity it needs to function as a sound base, while also giving transactional uses the liquidity they need.

Without a central authority and the force of law to safeguard the transactions, things come across as unsafe and risky. But that is exactly what Satoshi was thinking of when he designed the currency. Without the interference of agencies and central banks, as well as treasury departments that are typically held to political vagrancies, the Bitcoin market is undeterred by short-term and shortsighted manipulation.

This brings to question the records that need to be kept. There is no need for trust, and of course the trust that is typically handed to public officials to do the right thing is usually enforced by law. But all those effects come after the crime—when it's too late. So the Bitcoin system does not include trust as part of the system. Instead, it is all driven by records. The records of each transaction, from the first block of BTC to the last one traded a few minutes ago, are all preserved.

Where is it preserved? Each transaction of each bitcoin is stored on every node. A node is every computer that connects to the BTC network. It follows the P2P concept. In the P2P system (the peer-to-peer system), everyone who is on the network or using the service forms part of the group. In this case, every single node is a computer that is logged on to the system. On each computer, there is a client loaded and that opens a channel to the network making that computer a node.

The node also has the ability to store transaction information. Each and every transaction that is conducted leaves a record that is recorded in each node within a ledger. So there isn't a single repository of ledger information; there are hundreds and thousands stored across millions of computers worldwide. Erasing this ledger is simply not viable, and this record of every transaction is kept, which gives legitimacy to each and every subsequent transaction. That's because the record cannot be forged, unlike fiat currency. And it can't be printed to dilute the value.

The core of the cryptocurrency is the nodes that are within the system and the blockchain that keeps the ledger. It is what we call a distributed system, rather than a centralized system that the governments use to store value and manipulate it. This also makes it difficult for those who are speculating on currencies to get a good handle on the true value of the currency. It indeed increases the various dimensions of risk. That same risk is not present with cryptocurrencies.

So, here are the features of the cryptocurrency ecosystem you need to remember:

1. Individual coins that cannot be created
2. No centralized record of the coin
3. Distributed functionality
4. Market-driven
5. Can't be counterfeited
6. Easily transmittable

7. No records of transactions

Before we move on however, it is important to remember that a sound understanding of the functionality of the Bitcoin and the cryptocurrencies that were built on top of it, is a prerequisite for you to be able to learn how to effectively trade the currency.

Chapter 2

The Blockchain

How it works is the genius of the whole concept. Cryptography uses advanced mathematical equations to create unique strings of characters that can't be manufactured with hopes of it being deemed legitimate. There are two parts of the string of characters. The first part is unique to the unit of currency itself. So, imagine the dollar note in your pocket. If you examine it, you will see a serial number printed on it. That number is unique, and unless you manage to meticulously copy the design and printing techniques of the bill, you can't really do much with the serial number. But as we all know, notes are easily replicated these days. But comparing with, say, Bitcoin, what you have there is that serial number which is verified by the ownership of the coin.

When you open an account (anybody can open an account, which is called a wallet), only you have access to that wallet to be able to transfer money out of it. Anyone can transfer money to it, but only the owner can transfer money out of it. As soon as the money enters the wallet, it completely changes the entire serial number of the coin. So if Mr. A has a coin that he gives to Ms. B for whatever reason, the full serial number (string of characters) of that coin will be altered to represent the new owner of the coin.

We will look at the detailed keys and addresses in the relevant chapter of the book later. For now, just know that there are two parts to the full number. One is the number of the owner's account and the other is the string that identifies the actual coin. When Mr. A owns the coin, the coin takes on the identity of the coin and Mr. A's account. When he gives that coin to Ms. B, that coin changes its total string—shedding his account reference and taking on her account identity.

Technology

The beauty of the system is that now the account has ownership of money which cannot be stolen. Not like cash, where if you manage to take someone's cash, there is no way to prove that the cash is the robbery victim's. In cryptocurrencies, it is not even possible to make such a change of ownership—unless the device that held the account itself was compromised.

If you understand that concept, then the next question you should have in your mind is, "How does the coin change its string and who keeps track of it?"

Again, that's the genius of the system. Let me answer the second part of that question first. The infrastructure of the whole system is based on something called the blockchain. The blockchain is a framework that is made up of all the users of the system who are tied by a peer-to-peer framework. Remember Napster from about twenty years ago? It was a peer to peer system and a way to share music files. It was extremely popular, and all kinds of people came together to share almost anything you could think of in terms of music.

The way it was done was that you would put whatever you wanted to share into a folder on your computer and then download the Napster P2P software. This software would then go online and attach itself to a system where all the other peers were attached. And everyone kept track of what was in everyone else's folder. So if you wanted "Fly me to the Moon" by Frank Sinatra, and did a search, you would get a list of all the computers in the P2P network that had the file. You just had to choose which one you wanted to download it from—and when you clicked on it—a direct link between your computer and that computer was made and the files would transfer to you.

In that same overall conceptual structure, the blockchain is made up of individual account holders who leave their Bitcoin client running on their computer. When they join the network, the program will lock on to a number of other nodes (what you can think of as peers). They do not need to connect to all the existing nodes, but instead, a small number that is randomly chosen based on the wallet that you have and the settings you chose. What this connection to the other nodes does is extract all the data that is on the other node and keep an open channel with that node, checking for updates.

At the same time, there are other nodes that will connect to you that will feed off the information that you have about the client. They are not specifically looking to see how much BTC is located on your computer, but instead, they are looking to see if you have the most recent set of information. If you have information that is newer then what they have, they will copy that information and will also check if they're getting corroborating information from another node.

All this happens pretty quickly and every new transaction is recorded, then propagated through to all the nodes worldwide. Pretty soon, all the information about any particular transaction is available across all the nodes.

Using various algorithms, it is possible to trace the path of a coin. You'll stay in the know about when it was transferred, when it was kept, and when it has come online or if the account holding it has been active. But what you will not know is the identity of the person holding the coin. You will, however, know everything else. There is a significant amount of transparency in the system and a significant amount of security.

There are so many nodes in the world at this point that it is next to impossible to counterfeit the system or create records that are false.

Risks of the Blockchain

There are significant risks to the blockchain, regardless if it is the original one developed for BTC, or others that sprout from there. In the time BTC has been around, it has been declared comatose or dead more than one hundred and fifty times. But each time has seen BTC come back to life with increased vigor and vitality. Of those one hundred and fifty times that it has come close to being dead (actually, there have been more, but there are some cases which I don't consider legitimate enough to include in this book), some of those instances were from attempted theft, hacking, government intervention, and so on. But the popularity of the coin has come back every time.

Why? Because the underlying group that supports the market—not just those who have something to gain by virtue of its existence—but the unshakeable holders who rely on it and believe that fiat currencies are nearly on their last leg or deserve to be retired. There is a worldwide movement that sees Bitcoin and other altcoins as a way to liberalize and democratize the monetary system that is currently antiquated and controlled by the few at the expense of the many.

This brings us to the first risk of BTC and the fact that there is no prevalence in ideas that suggest that it could be shut down or ended anytime soon. However, it is still a risk nonetheless and you should make sure that you see it as such.

The primary risk and all the forms that it takes on, is if the ability of BTC to exist is extinguished. That's the main risk, and the individual risks that stem from that, serve to identify the areas that could be cause for concern.

The first one is that it could be shut down by a concerted effort of governments because they feel that it could undermine sovereign standing. That argument, although put forth in more ways than one, is a self-serving one. It's put forth by those who do not seem to understand that the only ones soon to be out of business with the eventual prevalence of cryptocurrencies, are the large banks. That's because they are close to the governments and thus have the shared perspective of cryptocurrencies. They tried it with BTC and failed numerous times. And the reason for that is, the people came back strong and hard in support of it. Also, there was no way to actually enforce whatever ban they wished on it—until and unless the internet itself could ever be censored and reigned in. Since I don't see that happening anytime soon, this argument of the end of BTC hardly holds water.

The next risk that you have to ask yourself about, has to do with the ability for sufficient nodes to keep track of the currency. We look at BTC and the blockchain as being the trunk of the system and the master ledger and record of the transactions that occur. When you can track each block and make a chain of them, the present standing of the currency coin takes on legitimacy. That entire blockchain is what gives legitimacy to the coin, and without that, it's become almost nothing. The fiat is given legitimacy by the government of the day, but that government is subject to political whims and fancies. Trust is also a subject that is negotiable in fiat currencies. But not so in cryptocurrencies. This legitimacy stems from an undying, unyielding record of each coin's existence and each transaction that it was a part of.

But the problem with this is that the transaction rates—especially with increased levels of trading—are making the blockchain grow by almost 20 gigabytes a year. And that is expected to increase. If each node is responsible for keeping a record of every transaction, then in just a few years there will be no one willing to host the full set of the ledger. As it is, most of the nodes that exist are lite ones, only recording transactions that are related to their own purse and transaction. If the file of the ledger gets so large that only a few people start to religiously maintain the ledger, it starts to give way to possible manipulation, if expertly done.

This is a risk, and the market has not priced that into the accounting just yet. One of the risks that you would have to anticipate in the next five years is the fact that the size of the ledger would be a problem. This risk is something that was mitigated by the use of the lite nodes, but could still be a problem in the future.

The next issue is the hacking of the system. The blockchain uses a 51% percent rule where if 51% of the ledgers on all nodes says that a particular transaction is correct, then that is the transaction that is recorded across all ledgers. In the event of a legitimate transaction, the 51% is achieved very quickly and in fact it goes all the way into the 90% range. The reason it doesn't get up to 100% is that not all nodes are always online to be able to record and verify from their particular node. However, if someone were to spend the time and resources and hack 51% of the nodes, then that would alter the security and usability of the BTC or cryptocurrency in general. So far there is very little chance for that to happen because the source codes for the wallets and the blockchain are peer-reviewed and open source. This means nothing is allowed on the blockchain until it has been verified and agreed upon. It is an efficient system and is world-wide, so the risk of this is also fairly small.

These are the bulk of the risks that could befall BTC and have an adverse impact on any long position that a trader may have open. But the thing you have to note is that the downside is low. Here is why. When any of these events happen, except for a concerted government shutdown, then the effect on the price would be gradual. It wouldn't become instantly worthless, but it will start to decline in value. In which case you could do one of two things. You could either short the coin or liquidate with minor losses from your long position.

Here is what you have to remember. There have been multiple events that have seen the critics declare the coin dead. Why? Because someone stole it or hacked it. But what they didn't see was that whatever the price drop following the hack or theft, was that the downside was low and the return was within a day or two. So the risk of these is not consequential, but gives rise to two trading opportunities—a short on the event and a long on the rebound. Many investors fail to see the value of shorts and leave cash and profits on the table. You should get familiar with shorts because that will help you ride the storm and the surge whenever they occur. They also make you more adaptable. Most people moan and groan when the market is bad. When the price is descending, if you can figure a short strategy, you would be happy whichever way the market moves.

Now we'll come back to the risks of the blockchain.

Most of the risk—aside from the possibility of a government clampdown, which is not something that can be totally discounted—is that you can see it coming, and the community will absorb the risk. If anything it is a trading opportunity.

Now, about the government clampdown. In the event it happens, the thing that will allow BTC to advance, is that there is no way of shutting down a P2P system. One that is fragmented and open-source can never be totally extinguished. The more the government clamps down, the more they will lose the ability to monitor and tax it. So in the end, they will see the light—as most governments have today—to be a promoter of the technology rather than being a detractor.

The risk that we have discussed here is both in terms of the individual currencies and to the blockchain in general. Generally, the blockchain is the concept of distributed recordkeeping and ledger updating that happens across all nodes. It is the common feature of all cryptocurrencies, and any small or minor changes from one version of a cryptocurrency—let's say from BTC to Ethereum—is superficial. It's not something that will completely alter the face of the blockchain that is developed for it.

The risk that occurs between each currency is comparative. The risk to the newer cryptocurrency is higher than an older one, assuming that the penetration rate is proportional to the time that it has been in existence. That risk is almost zero in Bitcoin compared to a currency that is, for example, created today.

Chapter 3

Understanding the Existing Cryptocurrencies

The original cryptocurrency was Bitcoin. It was the first cryptocurrency to capture the widest interest and the attention of the online world and the fiat currency world, as well as regulators and governments alike. It was the blockchain that gave rise to BTC, which became the template for all other blockchains and ledgers. Although improvements were made with other forks and iterations, BTC and its blockchain remain as the core and highlight of the cryptocurrency world. There are two general universes in cryptocurrencies. The first is the BTC universe, and the second is the altcoins universe.

The altcoins universe is populated by the likes of Bitcoin, Ethereum, Ripple, Dogecoin, and a number of smaller, recent cryptocurrencies. The first thing any astute reader would question is why there are so many different cryptocurrencies, how do they all enter the fray, and how do they affect the trading mechanisms?

Bitcoin

Because there is no central authority or sovereign behind any coin, and because there is no need to maintain large servers or incur a massive upfront cost, there really aren't many barriers to entry for anyone who wants to set up their own cryptocurrency.

In essence you could decide to setup your own currency today and be one of the more than one thousand different cryptocurrencies that make up the altcoin universe. Yes, there are more than one thousand cryptocurrencies in circulation today, ten times more than all the sovereign currencies of the world combined. Together they account for approximately $200 million worth of value. This may seem small, especially when you compare that to the world dollar-trading volume of about five trillion dollars, but you must take one more metric into consideration when you make comparisons like this. The dollar is the underlying and reserve currency of the world the instrument of consideration for more than 90% of all international trade. In fact, it's more than 90%—but not 100%—because the dominance of USD in the world is receding and countries like China are conducting global trade in the yuan. The dollar is also significantly older. Bitcoin is just a fraction in age compared to the dollar and it works completely differently.

There are two ways you can look at Bitcoin. The first is

to look at it from its currency profile. That is the part of it that you can purchase, sell, hold, trade, or use as a currency as payment for goods and services. The other side of that is the standardization of trust. Because of the blockchain that stands under it, Bitcoin is able to validate and verify not only purchases, but actual transactions. So it can be used in real-world situations as well. When a transaction is made and stored in the blockchain, the system-wide duplication of the event makes it unerasable and transparent. These are the reasons the advanced uses of Bitcoin have made it more compelling. Even with the outlawing of BTC in some countries around the world, the force in which it has come to the forefront is still something that cannot be overlooked or oversimplified.

Bitcoin carries with it brand value, and the fact that its algorithm and surrounding apps are stronger than the ones that have just entered the market. The inherent value in BTC is the immunity to fraud and the inability to alter historical transactions. It removes the trust issue from the equation and institutionalizes the transaction without sacrificing privacy. And the system has been proven time and again, which gives Bitcoin the edge over the other currencies. That edge is translated into the pricing of the Bitcoin in two dimensions.

The first, is that it's priced significantly higher than the other coins. Secondly, it is also more volatile in pricing than other coins. For traders, that means there are two advantages you can reap form this. The first is that your investment becomes more secure in liquidity, and secondly it means that there are more trading opportunities in any given day. So the idea of swing trading or day trading is rapidly becoming one of the avenues for wealth creation.

Here is a summary of Bitcoin that you should keep in mind when you embark on a trading strategy:

- Largest developer ecosystem
- First-mover advantage
- Highest liquidity
- Largest user base
- Proven security
- Store of value
- Far more accessible

These factors make it compelling to use BTC as the foundation of your trading strategy, although it does not have to be. But one thing for certain is you do need to understand the mechanics of BTC and the developments that follow, because its adoption, evolution and support will be a function of the health of the overall market. BTC has become the de facto bellwether of the cryptocurrency market and industry.

Altcoins

Altcoins have a strong and compelling story on their own. And as much as BTC is the default standard, cryptocurrency would not be as much of an industry if it were the only player. There are more than a thousand altcoins in the market today. The largest of these is Ethereum (which may change by the time of publication or soon after). Then there are the small, unheard-of altcoins that any entrepreneur can create with a proprietary blockchain.

You can create your own altcoin and name it whatever you want. You do not need authorization or a license, and you can use it as the medium of exchange for anything. You could even have a pizza place and say that the only currency you accept is XCoin (random name), and sell XCoin on the side so that people have a way to access it. At the end of the day the coin is merely a token that represents an intention and a value—the intention to transact, and a value to assign to the consideration.

Using those same ideas, thousands of altcoins have come onto the market with differing concepts based on the same framework. This has been a solid development, and instead of taking away from BTC, it has enhanced the entire market as a whole—which has pushed BTC higher. How does that benefit you? It gives you an opportunity to trade in 0the market and ride the fluctuations, appreciation, evolution and advancement of each entrant into the market.

There is a balance that you have to find between the coin that is entering the market versus the coin that has been in the market for some time. You need to find the one that gives you the sense of security and of return before embarking on a transaction. That's why, when you trade, you try to limit yourself to just one, or a small handful of coin(s) as opposed to one particular coin. Typically, you would want to become an expert at, say, BTC vs Ethereum. Or, BTC vs Monero. The actual pairing doesn't matter, but you must keep up on your knowledge of the coins, with a minimum of two pairs. Even if you are a day trader, there must be a certain underlying knowledge of the currency in which you're expert, so that you can make decisions quickly and keep a better eye out for opportunities when they arise.

Once you get an initial idea of the pairing you want to be an expert in, and gather the knowledge to get you to that level, you also want to be able to take on at least two more pairs so that you have four pairs in total. Those four pairs include BTC as part of two of the pairs, but not the other two. So let's say you have BTC vs Ethereum, and BTC vs Litecoin. The third pair could be something like Ethereum vs Bitcoin and Ethereum vs Ripple. These choices are not meant to advise you on the pairs you should concentrate on; they are merely illustrations of what your basket of knowledge should cover. In these four pairs of currencies, you only need to keep up on four individual currencies. This will make it easier to understand the movement of one against the others. It is also an efficient way to get to understand your area of expertise. As an ex-trader myself, I used to focus on USD vs GBP, USD vs Yen, and GBP vs Yen. That opened up a larger set of opportunities in cross trading and arbitrage.

Even though the cryptocurrency market is not as liquid and rapid as the currency market, it will eventually get to that point; and your practice with it will pay off in the long term. But aside from that long-term benefit, the ability to focus on only three or four currencies and have a larger set of opportunities on the cross-trading circuit, gives you a better bang for your buck.

Even though I am jumping ahead to strategies which are found later in the book, let me give you an example of what I am talking about so you can see the benefit of focusing on three or four currencies and cross trading them.

There are days when you will not be able to get a decent price of opportunity to execute a trade that is worth the effort or the risk in one counter, and so you will want to shift your focus to something away from it. So if BTC is having a slow day, then you will not find much opportunity on BTC vs Ethereum or BTC vs Litecoin. But you may find some activity in Litecoin vs Ripple, and if you have a diverse skill set in the counters you will be able to take advantage of that day. When you are a day trader or a swing trader the amount of wealth you create is directly proportional to the number of quality trades you make, the ratio of wins to loses and the number of times you enter and exit the market. So when you have more opportunities to access the market by understanding the nature of more pairs of currency and how they move against each other and how they value themselves, you will find more opportunities to make a trade and to be rewarded.

Here are the ten top alternative coins that are on the market. Remember there are more than a thousand; these are just ten of the ones that you may want to trade. The others are not ready to be a trading opportunity and won't be, until they have a compelling reason to be a profit potential.

1. Ethereum
2. Ripple
3. Litecoin
4. Dash
5. NEM
6. Ethereum Classic

7. Monero

8. Zcash

9. Decred

10. PIVX

Tradability is not just about price. Something can be extremely valuable in price, but if it is not volatile, it presents no opportunity to trade. If it is too volatile, it presents reduced security to trade. So a balance must exist in the level of volatility and the liquidity of the market. The best way to trade a coin is if there is sufficient liquidity and heavy volatility.

Why? Because volatility could happen because of the lack of liquidity. In this event, if you entered the market and traded the coin, you may not be able to get back out. And the trade may be a floating profit for a moment but turn to a loss by the time you find a counterparty to execute the transaction. In the case of liquidity, the more popular and the higher the user base is, the more liquid the market and the lesser chance there is for the currency to have wild price gyrations due to momentary purchases and sales. This is one of the reasons Bitcoin is more expensive than other altcoins—because it provides better liquidity. This liquidity premium prices in the effect of immediate entry and immediate exit so that the price reflected on your trading screen is very close to the price you get when you execute a trade.

Liquidity

There is no doubt that there is a rush to create the next bitcoin or to surpass it. There have been millions, even hundreds of millions invested by old-currency investors to be able to dominate the future of cryptocurrency. And as a result, you get figments of bits and bytes that can rack up values of over $5000, like Monero soon after its launch. There is a specific kind of risk profile attached to each cryptocurrency and that risk profile is a function of its liquidity.

The liquidity of an asset determines its overall value, price, strength, and most of all, the belief fans and followers have in a particular asset. Look at gold as an example. It is merely metal and has some properties that are interesting, but it is just another item on the periodic table. It is the allure human beings attach to it that makes it valuable. The same goes for the value of cryptocurrencies. It is the value that humans place on it by their actions. The more they buy it, the more they trade it—and the more they use it as part of their transaction, the more it becomes the de facto currency. Much in the way the USD became around the world. But everything starts somewhere.

To be good at cryptocurrency, you have to understand its popularity, and many of the factors that lead to it need to be internalized by close study. The best traders instinctively understand the way the currencies work and how they respond to an event on the market.

How then to choose currencies to trade? Whether it is purely cryptocurrency—which means it's one cryptocurrency for another and vice versa, or it is from a fiat to a cryptocurrency—the thing you have to do to pick the currency with the most profitable pairing. That's the one that gives you the ability to ride the transactions effectively. There is a benefit to working with volatile coins if you are able to find the necessary liquidity to enter and exit the market. Liquidity is your most important factor.

But what about liquidity in coins that have not been established for long? Look at BTC vs Monero. The value of BTC still outweighs the newbie, but that still allows for a trading strategy. So in this case what you have is a holding of BTC as the core, and parts of it are used to buy into Monero. That way you benefit from the security of the older coin, and the potential appreciation of the young coin. This is a strong strategy and you should use it in part of your trading endeavors.

Typically, that would look like this:

Open an account with a measure of BTC. In this case let me use 100 BTC as an example, assuming you have 200 BTC as your starting point (remember the 50% rule). Then use the BTC to divide between a couple of trading strategies. The first is to use BTC to buy the currency that you have chosen, let's say Monero. The first pocket of purchase would be about 10% of the total investment value. This is your long-term strategy. You buy this as a value investor. Next you place about 15% of your BTC into Monero at a price point that is indicative of a swing trade (with a horizon measured in days). Finally, you use 25% of your funds to buy Monero that's to trade on a daily basis.

This is one strategy that you can use. If you are bullish on Monero, then you would employ this strategy. If you are bearish on Monero, then you should short the coin directly from your dollar account. Remember that each coin will move in a slightly different way to its counterpart, because their relative strength varies.

Here is what I mean.

Imagine if you calculate that Monero is about to appreciate in the long term, then you follow the strategy above. That way, you are making money on the long-term appreciation, short-term appreciation, and the short-term retracement. You're also making money on the fluctuations that happen on a daily basis. There is a possibility that at any one moment you may be long Monero and short Monero on different horizons—and making money on both. That's the ideal situation and that is what you want to do with the trading strategies.

Many of my customers ask me why they should be

selling a coin when they are bullish, long-term. The answer is that you want to take advantage of the minor fluctuations which happen over the course of an upswing, or a downswing for that matter. There are numerous opportunities in any trade and you want to maximize your entry and exit so that you can make a larger return on the same amount of investment.

This is called "scalping the market,' and there is an absolute fortune to be made from this if you get used to the nature of the currency that you are specializing in. This is why liquidity is so important. If you are in a liquid market you can convert that liquidity into profitability. You can take advantage of every uptick and downturn and still make money on the long stretches of appreciation.

But as a trader, you cannot just depend on the fundamentals to make a trade. Relying on fundamentals is only about 25% of your trading strategy. The other 75% consists of a strategy that is based on statistics and historical trends.

Volatility

In parallel with volatility you have liquidity. This is a harder measure of a currency's investability. A volatile currency is a function of a number of factors. There are currencies that are inherently volatile because of a number of their underlying features. Then there are assets that are volatile because they do not have many owners or buyers and sellers, so that whenever someone wants to make a purchase and the come out the of the market the place a sell order and it becomes a buyer's market.

A buyer's market means that the buyer can wait out the seller till he lowers the offer because he is in a rush to liquidate his asset. Or it could be the other way around. When the buyer comes to the market and wants to buy, he has to raise the price behind what the last trade made so that the seller is enticed to come to market and part with his lot. This creates an artificial pricing scenario, but really it is the premium for the seller to sell when he was not planning to, or the buyer to buy when he was not planning to. The price difference between the buyer and the seller gets further apart the thinner the market gets. When the market thins out, any trade is going to be largely different from the trade before. With a thin market, it resembles more of an individual seller or an individual buyer and loses all form of predictability. And with that, the pricing gets skewed.

When you choose a currency, you want to make sure—as we have repeatedly said in this book—that the currency needs to have a sound level of liquidity so that the volatility is not severe. Liquidity is not just so you can get out when you want, it is also about being able to get a good handle on the price that is showing up on the screen. The thing about volatile markets being like they are because of low volumes, is that it tends to mess with the auto-trading systems and the indicators for buy and sell points. The bottom line that these metrics use is that there is always a buyer and a seller available so that the pricing is as fluid as possible.

Stay away from illiquid markets unless you are in it for the long-term gain of doing something strategic with the currency. If you have the appetite and the ability to hold or lose all your investment, then it is a good strategy as long as you have the right goal. But if you are just a trader looking for a return on investment, stay away from illiquid markets.

Chapter 4

How to Purchase Cryptocurrencies

Purchase Risk

There is a risk that you face when you purchase coins from unknown sellers. Especially if you are doing it with fiats. There is a chance that once you pay for it, they do not transfer the coin. So that is a risk that you must contend with when you find a broker to purchase the coin and when you start trading. There are two strategies to overcome this. The first is that you can buy the coin from a reputable seller. That way you deposit the funds and the coins are certain to reach your wallet. From there you transfer the contents of your wallet to a hosted wallet with the trading broker. Then you can make whatever trades you want. That takes out the purchase risk. But you still have one more risk when it comes to trading.

What most people do not understand is that there are a number of trading houses out there that operate on the fringe. What they do is take your coin (be it BTC, Ethereum or Monero) and they then make the purchase on the market or you. Sometimes, they do not make the actual purchase and they just record the ledger as the purchase being made. This happens a lot in all brokerages because one of the ways traditional financial brokers work is that they buy popular assets in bulk, keep some for themselves and their discretionary accounts, then apportion the rest out to other customers who also bought them.

That's fine—when the system works. But there are times when the broker may be so risk-tolerant and fancy with the accounting that they do not actually make the purchase, but instead take the opposite bet from you. In the event you lose money, they are covered; they make the money that you lose. but in the event you win, they typically dip into their own reserves to pay you. They are betting that you will lose more often than you win. But the problem arises when they pick wrongly, and a large volume of their clients profit from this trade they made the wrong choice on. That results in a situation where they are not able to cover the position and they wind up.

This risk happens with new brokerages trying to make in big in short order. When it comes to brokerages, you have to be smart. Either chose a brokerage that has been around for a while, or one that provides you with the safeguards you demand.

Purchase and counterparty risk are very real in all trading environments and the same applies to a larger degree in online cryptocurrency brokerages. The rule of thumb is that if you lose your money to a brokerage that goes upside down, you will never get your money. Be comfortable with that fact and place only what you need to with them at any one time. If you are going to be a heavy trader, then you need to set up the necessary infrastructure to cold-store the coins that are not in play. And since you should only be investing half of your coins (at the most) at any one time, or investing at two different brokerages who aren't related, then it becomes worth your time to set up an off-line storage that is air-gapped and secured.

Chapter 5

How to Invest in Cryptocurrencies

Investing in cryptocurrencies is a fairly easy process. It is not as stringently regulated as the stock market or even the Forex market. That lack of regulation gives it an air of the old west, and this is where you make your money without the intrusion and interference of burgeoning regulation.

The one thing that you are not going to read in any other book, and that is exactly why you will find it only here, is that investing in coin should be done in a way that is different from investing in equity or currency. It should also be done differently from how you would invest in real estate or precious metals, or even the futures market. Why? Because cryptocurrencies are a different kind of asset. They are not physical assets but conceptual ones.

These cryptocurrency assets straddle the conventional boundaries of tradable items. On the one hand, there is no underlying asset. If you look at a stock price, there is an underlying share certificate that has a derived value which is attached to the earnings of a company. If you look at the price of land, there is a piece of tangible tract of earth that is the basis of the price and whatever the present value is of the future rental of that land. But in a cryptographic coin, there is nothing underlying it at the time of creation. It is merely a code that is unbreakable, unduplicable and has value only when it is owned. If the owner erases it, it ceases to exist and to be of any value. The value of a coin is not that it is coin; it is of value because it can morph into the value of whatever it is buying and what its buyer says it is. That is unlike fiat currency, which is used by the government to create a value they deem it is worth.

So when you trade it, how do you set up your frame of mind? You think of cryptocurrencies as if they are the sentiment of the human collective. That's the trick that no one seems to get. You cannot do a DCF, as you would the price of stock. This means that you couldn't possibly get an intrinsic value of a note of BTC, because the coin itself has no future value, and thus you could not discount it back as you would do a unit of stock. You also could not look at the demand of the coin in terms of M1 or M2 in money supply or the balance of trade between currencies, to understand what the underlying asset looks like. So that is why you have to look at it as the mirror of the human psyche and the potential that it could eventually become.

When you do this, you start to see the pattern of the masses—including its irrational exuberance of riding the market up to near $20,000 and then falling back on fears of it being too high—a characteristic that is all too human. From a philosophical perspective, all tradable assets inherently shadow the collective mindset of civilization.

This is why one should trade the units of currency on pure math, because the value of the currency can be statistically approximated when using trading algorithms. That is indeed a strategy that you can use when you are scalping.

Buying Low, Selling High

This is obviously the point, but I need to state the obvious just in case there is some sort of confusion—especially since we will be talking about shorting the market in a short while. Buying (or going long) is something that you talk about when you are discussing the base currency that you are purchasing. If you talk about buying BTC, then there is an underlying assumption that you are using fiat currency to purchase it. And to drill down further, you will be thought to say that you are buying BTC with dollars. When you address the purchase of the currency, you are talking about buying the asset low and selling the asset high. The reason this gets confusing is if you buy the BTC to be able to invest in something else, then you are actually using BTC as the replacement of cash. So you have a compounded profit calculation to make, depending on which asset you are using.

Remember that you need to apportion your holdings and keep track of your base asset. In most cases that would be your BTC holding, or your USD holding. For now, those would be the two that I would consider as my base asset.

There are three considerations that you want to make when thinking about your base asset. The first of course, is liquidity. The second is appreciation or stability potential, and naturally the third is volatility. These are the same three that you need to consider when you are looking at investment vehicles. The reason why it is most important is that you have to ask yourself what your long-term strategy is. Is this for your kids' college in twenty years, or is this your retirement nest egg? Or is the path you chose strictly to make supernatural profits?

The answer to this simple question will determine how you want to apportion your holding in cryptocurrencies and which asset you want to hold as the base asset. The reason most people use the dollar as the base asset is because most people will plan on converting back to the dollar for use later on. This is changing slowly, as even colleges are allowing tuition to be paid for with BTC. For that reason it has been measured that the trade volatility and the price stability of BTC have positively advanced. But do not misunderstand. That does not mean that BTC has no trading potential, because it is easy to think that fluctuations give rise to profit potential. That is only true when you are betting on the change in value that occurs on a day-to-day basis or shorter. You can scalp effectively with this strategy, but that is not the only thing you want to do when it comes to cryptocurrency trading. You also what to take advantage of the large moves.

So what we have established is that chasing the base currency is an important consideration in the process to invest and trade in cryptocurrencies.

The best way you can manage this is to then think about your long-term objectives and apply the same portfolio management practices that we see in other assets. As far as cryptocurrencies are concerned then, you need to apply the smallest margin of your total portfolio when you are T-10 (ten years before retirement). That means when you are ten years away from your retirement, it is the point that you will have the smallest holding of cryptocurrencies in your portfolio. When you reach that threshold, you should no longer invest in cryptocurrencies because you are not going to be able to recover losses that may occur.

It really is a matter of perspective when it comes to buying and selling. Buying BTC using the dollar is the same thing as selling the dollar in favor of BTC. The trade is coupled that way. It's hard to fathom that if you are new to currency trading, because usually you are using currency to buy paper assets like stocks or futures. But in this case you are using currency to buy a currency. Buying one means you are selling the other.

Selling Low, Buying High

This is the exact scenario that we saw earlier, and I have to include this so that we get a proper understanding of the mechanism. There are a lot of you out there that do not have experience with currency trading and so it becomes a little hazy when we talk about selling short or buying long.

If you understand the previous section on buying low and selling high, then the scenario in this section becomes a little perplexing. This scenario is not about an endeavor to lose money. It is just the exact reverse of the above scenario. When you buy BTC using Monero, for instance, you may be buying BTC, but you are actually selling Monero in the process. For every transaction, you are doing both acts of buying and selling, simultaneously. This is something you have to get used to so that when you start trading and when we get into the deeper aspect of trading in the upcoming books, you will be able to freely move from one currency to the next without missing a beat.

Hedging

What's a hedge? Well in simple terms, a hedge is a fence

that sits between two sides. In a trade, a hedge is a position that straddles both sides. That means for every lot you transact to buy of one asset, you short the asset in a parallel transaction.

Purchase and liquidation orders are unique in cryptocurrency transactions. In this kind of transaction, you need to understand the peculiarity of buying and selling. When you buy an asset, then sell its corresponding asset, it is normally taken that you have rid yourself of the asset. But that is not effectively true in trading. To sell the asset that you own, you need to liquidate or sell-liquidate it. When you specify that it is a sell-liquidate, then the broker will sell the asset that you already have purchased and you will realize the profit or the loss that you incur in that transaction. However, you also have the option to sell new. That means in addition to what you already own, you make a parallel transaction where you sell the asset by shorting it. This will be taken as a new transaction.

What is the end effect of doing his? Well, what happens is that it locks your losses into a specific range rather than allowing them to free fall. Let's take for example a hypothetical cryptocurrency (let's call it X2) against BTC. Let's say you bought X2 for $1 BTC. So that means at the time of purchase X2 was at par with BTC.

The investor's purchase of X2 was with the calculation the X2 would appreciate in value and thereby provide an opportunity for a cash-out at approximately 1.2

BTC. But in an unfortunate turn of events, X2 got hacked and the currency instead started to depreciate. If the price had already gone to 0.9 BTC, that was already a 10% loss. So instead of cutting the position and materializing the loss, the investor opened a new trading position and shorted X2.

Keep in mind, on one hand he has X2 as an asset. On the other hand, he has sold X2 with a promise to deliver it at another time in the future. Once he has sold at 0.9, and has a buy position at 1.0, any movement of the price will see that he has a trailing loss of 0.1 BTC. Let's say the price goes to 0.7 BTC. His 1.0 buy position would be losing 0.3 BTC, while his 0.9 Short position would be making 0.2 BTC. His net would still be a floating loss of 0.1 BTC.

At the same time, if the price were to fall further and go to 0.5 BTC, then his original buy would be losing 0.5 BTC while his short would be earning 0.4 BTC. Again, his floating loss would be 0.1 BTC. In the event he gives up, he can cash them both in at the same time and come out with the realized loss of 0.1 BTC.

What would be the point of that? The point is, the event that occurred while the trend was for an appreciation, temporarily knocked of the ascent and took a momentary dive. At that time, hedging that position would safeguard against unmitigated losses and also creating a downside trade. When the price reached 0.6 BTC on the way back up, or a 40% loss on the 1.0

BTC long position, the investor could cut 0.9 short position and reap the profit of 0.3 BTC while still facing a .4 BTC floating loss. The equity position on that account would still be -0.1 BTC.

But the moment the rebound consolidates and the price returns to 0.7 BTC, the long position would be losing 0.3 BTC, with the accumulated profit of 0.3 BTC resulting in a net equity of 0 BTC. By the time the rate returns to 1 BTC (the original long-position entry point) and the floating loss is zero, the profit reaped by the short position would be free and the equity would now be 0.3 BTC.

When the price goes to 1.2 BTC as predicted, the profit from the long of 0.2 BTC, plus the profit from the hedging short of 0.3 BTC, result in a 50% return on investment. This, as opposed to a loss of 0.1 BTC or 10%, as the investor had liquidated his position when the market turned on him for a moment.

This is the value of a hedge, especially in the period of a crisis.

Chapter 6

Day Trading vs Long-Term Investing

There are a number of factors that determine the length of time you should open yourself to a trade. It is not only dependent on how comfortable you are, or what your schedule is like. If you are a part-time day trader versus a full-time one, that will significantly have an impact on the choices you have. On the other hand, the amount of capital you have and the length of time you have been in the market will also have an impact on this decision. That applies to any market.

For the cryptocurrency markets there are a number of other considerations that you have to take into account. Remember that you need to view them as a hybrid of assets rather than looking at them purely as a currency asset. There are peculiarities that lend themselves to the cryptocurrencies, especially to BTC, and there are other peculiarities that apply to the altcoins. We will look at them in turn.

Deciding if you want to trade something on a daily basis or if you want to hold it for some time needs the consideration of two areas. The first consideration is instrument-specific, and you need to understand the asset which you intend on trading. The second is the traders infrastructure. We will cover both of these as we go forward.

When you look at the asset, you need to see if it is going to provide you with sufficient opportunity balanced with security, to be able to make trades that result in a positive outcome and a higher than normal rate of return on investment.

Why do you need a higher than normal return on investment? As profitable as cryptocurrencies are, they are not representative of anything tangible. They represent the intangible state of relations between two parties, but when that is stopped or confiscated, the asset itself is non-existent. Imagine a barter trade. If I trade a herd of cows for a plot of land, at the end of the day there are tangible elements involved in the deal. If I advance the financial aspect of that example and trade a herd of cows for cash, I have one tangible asset and one financial asset. If I convert one currency for another (sovereign) I have two financial assets that are backed by the promise of a sovereign government to honor the value that is printed on the note. But in the case of cryptocurrencies, the value of the coin is purely and totally driven by market forces. It is the most elegant creative use of math and computing technology to date, but in the event of a problem, the worth of a coin vanishes into thin air.

But as we all know, risk is a comparative matter. There

is also the risk that the dollar depreciates and there is a risk that catastrophic events take place that diminish the value of the currency's underlying economy. We can't keep looking at every single factor because that would be too many points of data to contend with. So, we create rules of thumb to make sure that we categorize certain issues as the underlying fundamentals. Then we use dynamic issues on top of that to determine the swing or day trades.

Let me give you an example of this in a way that makes more obvious sense. Imagine if the North Korean currency was available to be traded. Your instinct will tell you right off the bat that you shouldn't put all your assets in that one basket—if you were willing to put any in at all. The reason is that the risk of depreciation is high and you instinctively get that—there is no stability in what the government may or may not do. With that kind of instability comes the issue of insufficient liquidity risk and volatility issues as well.

The other thing about risk is that you may even have asset manipulation. This is one thing that most investors forget to consider when they choose an investment asset.

Whether it is the dollar, the won, or a cryptocurrency like Monero, you need to ask yourself who has the ability to manipulate it? You may think that currencies can't be manipulated, but you would be surprised. Do you think that the Chinese yuan is at the level that it is naturally supposed to be because of market forces? Or do you realize that it is kept there in what is called a managed float? A managed float is when the government of the country decides the currency that is pegged to that particular rate is beneficial for the export and import ability of the country. A number of Asian countries do that to make it more conducive for Western countries to buy products at a cheaper rate. This creates a strong imbalance of trade, tilted by the artificially pegged currency—which is flat-out currency manipulation.

In that example, the country or sovereign that issues the currency is large enough to be able to absorb the downward pressure of the sellers. Is this not manipulation?

It is. We don't have this in cryptocurrencies, as there is no single entity like a sovereign government that needs to undertake an artificial pegging of the BTC.

You have to look at these factors and understand if indeed you are willing to stay in the market over the long term to be able to take that kind of manipulation. In the short term, that is ok, but in the long term, that could be a potentially huge risk for your trading assets.

Let's put it this way. I wouldn't event put a large portion of my investable assets in any currency, let alone cryptocurrencies. Thinking about that, the profit potential of the cryptocurrencies are twofold. Firstly, they offer fluctuations that can be traded in both directions. Secondly, they offer longer time appreciation because they are in the various stages of adoption. As more people come on board and use them, the more there is a demand pull pricing of the coin. That will eventually taper off and the pricing will stabilize. You are already seeing signs of that with BTC. Irrational exuberance aside, the factor that carries it to $17,000 is not one that can be sustainable and traded as a long-term investment, but the daily upticks and the retracement provide daily profit potential.

That is just for BTC. The longest that you can hold this for is between a day and a week to ten days at the most. You should only scalp, day trade or swing trade BTC.

As for altcoins, you should stick to scalping and day trading only. All the altcoins that are in existence have indeed learnt from the structural experience of BTC but they are still fairly young and not as established as BTC. There are still cases of hacking that affect the upward price moves, there are still thefts to consider and there are still adoption issues that have yet to be resolved. A day trading philosophy on the altcoins - all coins other than BTC, is worth the effort because you can still continue to trade them constantly on the way up, yet keep your money safe while you retire from the market at the end of the day.

Your mindset should be informed by the following when considering your Bitcoin trading strategy:

Bitcoin

- Not backed by an underlying asset
- Liquid
- Volatile (4 - 8% price fluctuations per day)

These three factors pull your decision on BTC towards a decision of being a day or swing trade candidate, because even if you keep to a twelve-hour trade day, you will be able to make more on the rise and retracements than you would if you just bought a position and left it there.

Let me illustrate.

I will use a typical cryptocurrency trading day without any major market fundamental news as an example.

Let's say that the price of the coin at opening is 10000 per USD. (I use round numbers to make the calculations simpler to communicate). During the day, the coin goes from 10000 to 10800. That is an 8% appreciation, but not all of us are accurate enough to enter right at the bottom and get out right at the top. But for illustration's sake, let's say that that's the maximum that you are able to make—if you do one trade.

But that's not how you play this game. You are in cryptocurrency trading because of its volatility and liquidity and you want to use the time to buy on its upswing and get out when it turns to retrace. Here is what that would look like.

If you enter the market at 10050 and it goes up to 10200 and starts to fall back, when you see it retrace, you liquidate your first position of the day at, let's say, 10180. That's a $130 profit.

Now let it find the bottom of the retracement and wait till it turns around and starts back up. You hop back on again, and let's say it goes all the way back to 10000 and starts turning back up. You can catch it at 10020 and ride it up again where it turns at 10200—and you get out—let's say at 10200 again. Then you let it fall back and it now goes on to 10150 before turning back up where you catch it at 10170.

You ride that wave all the way up and it goes to 10330 before it starts stepping back, and you get out at 10300 while it goes back to 10240 and turns back around for the next ascent. You get in at 10280 and ride it to 10430, where it starts to turn around. Then you get the signal to jump at 10400 while it goes back down to 10340—before it turns back up and you catch it at 10360.

At this point you ride it until the next crest. That happens at 10480 and you liquidate at 10460. It continues to 10400 before turning back, and you catch it at 10420 and ride it to 10550. In the same fashion, you make the next few trades. In at 10530, out at 10650; in at 10630, out at 10750; in at 10730, and out at 10800.

In total, here is how you made out:

In the first trade you made 180 (10020 to 10200); in the second trade you made, it was a repeat of the first at 180 (10020 to 10200); 130 (10170 to 10300); 150 (10280 to 10430); 100 (10360 to 10460); 130 (10420 to 10550); 120(10530 to 10650); 120(10630 to 10750), and 70(10730 to 10800). In total that is $1180. So instead of making just $800 on the one run, you took advantage of the nature of the cryptocurrency's ascent and caught it as it retraced and rallied again. And yet, that is still not the optimal outcome. There is still one better way. What you can do is to attack the market on its retracements *down* as well.

Here is what that looks like:

When you get to the top of the run where you liquidate your original long position, you enter a short order simultaneously. When you do this, you will be able to scalp the market on its retracements.

Here's what now happens:

All you need to do is add an additional order ticket to your trades. When you started, you entered at 10020 and liquidated at 10200. That liquidation order should be accompanied by a new sell order at 10200. You liquidate that position when you enter the market to make your next buy position is at 10020. So as you buy in at 10020, you also liquidate that short sale you made earlier.

This will bring you 180 for the short sale profit. Next you make your buy as above, at 10020 and liquidate at 10200. Here your liquidation order is accompanied by a new sell order for 10200. This gives you another 180 in profit. In the same way, you now keep shorting every time you liquidate your buy position.

And this is what that looks like:

First trade short position will yield 180 (10200 to 10020). Second trade will yield another 180 (10200 to 10020). The third trade will be at 10200 and liquidated at 10170, for a profit of 30. The fourth trade goes in at 10300 and out at 10280 for 20. The fifth trade goes in at 10430 and out at 10360 for a profit of 130. The sixth trade goes in at 10460 and out at 10420 for 40 in profit. The next is 10550 and out at 10530 for 20 in profit. Next it is 10650 and out to 10630 for 20. Then it is 10750 and out at 10730 for another 20. In total that's $640 in additional profit instead of just buying on upswings.

In total your long positions give you $1180 while the shorts will give you $640 for a total of $1820.

Look at the difference that the opportunity of a cryptocurrency with a volatile and liquid market gives you. You have the ability to profit as it swings upwards, plus you have the potential to make money on the swing down. In total, you make $1820 instead of only $600 from doing just one trade.

If you are thinking this entails a lot of work, well that is the reason there is software which you can use to purchase and sell according to your patterns and instructions. So, the strategy that I just gave you can be easily entered into a trading program and executed. When it is the software, it will follow the entry and exit points that you specify, then execute the trades. You just have to monitor it and make sure that it does not do anything that gets you in trouble. I keep all my programs in such a way that they suggest the next trade and wait for my approval and only execute if I approve the suggestions. If the timing between the suggestion and my approval is too long and the opportunity lapses, my software recalculates and gives me a new option.

I will give you a few more options that you can try in your strategies later on in the book. They'll give you an introduction on how to get your own trading strategies up and running, or at least give you an idea of how to go about doing it. Trading cryptocurrencies gives you a tremendous amount of freedom and room for creativity to be able to execute optimized trades like the one we just saw.

Chapter 7

Investing Strategies

There are three trading strategies that I will get into in this book, and leave the rest for the upcoming books in this series. We'll peep at a fourth one here as well, but it's a more advanced strategy, requiring a deep dive into a new book!

No, it's not so that you will buy more books, but for now, I am assuming you are just getting in and getting your feet wet. If I were to burden you with any more than these three—plus the one in the last chapter—it would be all too overwhelming.

So with that in mind, what I want to do in this second half of the book is hunker down and focus on the strategies that you can get up and running right now, to get yourself familiar with the nature of how bitcoin and the other altcoins work. I also want to say one additional thing before we get underway. There are thousands of coins out there and at least a dozen that are prominent, but the king of all the coins, and the market leader, is BTC.

We know this much. But what you should also know is that you should limit your exposure to the coins that are within this top dozen. Even though there are thousands of coins in the market, you shouldn't be spreading yourself so thin as to try to trade them all. You will end up overwhelming yourself and making lots of mistakes that will cost you time and money.

In a market where every minute sees a fluctuation and every fluctuation can potentially make you a profit, you should keep yourself in the loop at all times and know exactly what you are doing on a daily and weekly basis. I will show you how to keep your focus in the section on mindset in the later part of the chapter.

Buy the Dips

Your first strategy is to always buy on a dip. Do not buy into an already established rally. If you stop and watch the market for the first time, you will note that the market always charges forward, then re-traces, and then charges forward again. It is the nature of all financial markets—especially the currency markets, and more so, the cryptocurrency market.

If you buy at the moment you see the ascent, then you are in for heartache when it takes a dip, and it could demoralize you. So wait for the dip—and when it dips—buy it on the dip after it turns to rally again. That way, once it turns around, you have a longer stretch to run with. Let the dips be the indicator for you to pull the trigger.

The same goes for when you short—wait for the dip. In

this case, what the dip in the descent means is that it backs off on the way down for a minute—and that's when you catch it. If you look at the example in the last chapter, you will see that the same trade for the day was made when the currency started an upswing. When it crested and started coming back, is when it was liquidated. The thing to note is to never try to catch it at its peak. You may get lucky once in a while. You may get in at rock bottom and get out at the apex, but that is not going to be the way all the time and not what you should expect. Both apex and pit have a specific purpose, and that purpose is not for you to harvest or liquidate, but to prepare for the next move. Those are your trigger points.

On Balance Volume

The next strategy that you want to get familiar with is to understand that the majority of BTC are traded on a day-to-day basis. You want to be able to understand where the big money and smart money are going, and follow that. To do that, what you want to do is look at the flow of money, and the OBV indicator does exactly that.

The OBV indicator can be found in some of the better brokerages that offer MT4 platforms. This is what you want. The thing that you don't really want to do is trade blind and manually. You want to be able to program as much of the trades as possible with sufficient downside risk mitigated, and sufficient upside potential capitalized. In the OBV indicator, it shows you the amount of money flowing into a particular counter. In the case of BTC, if you see that the coin is trending up while your OBV indicator is trending down, then it means that the money is coming out of the market and the rally isn't sustainable. You can use this as a guide to get ready to short the market or liquidate a long position that you took earlier in the cycle.

On the other hand, if the price of the coin is descending but the OBV indicator starts turning up, then you are looking to make a buy sometime soon, because the money is starting to flow into the counter.

Here is how you can use that with buying the dip and with the strategy of short the down and long the ups that you saw in the previous chapter. You initiate the trade in the morning and program the software to buy purely on the dips, with the first warning coming from the OBV before initiating a position. You also place a stop loss in the system so that your downside is mitigated. Then too, you can place a conditional order so that instead of a stop loss it is a hedge like we talked about two chapters ago.

Arbitrage

The one strategy that you should absolutely follow

when you are new to the game is that of arbitrage. If you are not doing cross-currency arbitrage, you are leaving money on the table and you will have no excuse because we told you so right here!

Here is how you look at arbitrage:

Remember we told you that you should follow at least two to three currencies, with one of them being BTC? Remember to also follow the dollar and watch how it, and the other currencies in your knowledge basket are doing at all times. You have to know that the markets are all connected and that you need to keep an eye on that connection.

The way you set this up is to have your program monitor the three or four currencies that you are trading, and also look at the cross-currency quotes. So imagine we have Currency A, Currency B, C and D. Currency A is the base currency and quoted against the dollar, and B, C and D are also quoted against the dollar.

For now, let's say that Currency A is quoted as 1:1 against the dollar—that means you get one unit of A for every dollar. Then you have Currency B trading at $2, Currency C trading at $3 and Currency D trading at $4.

Let's just take two currencies at this point to make the example simple. If Currency A is trading at par with the dollar, then it should also be trading at units against Currency B. But if you watch the market closely, every few hours you will get a scenario where Currency A and B, instead of being 1:2, will be at 1:1.9. Assuming that the base currencies of Currency A and the dollar and Currency B and the dollar are still on par and 1:2 respectively, there is a price mismatch between A and B.

This is an opportunity for you to buy up B using A, and then using A to buy the dollar. That will yield you an immediate 0.1 profit. There is almost zero market risk in that as long as you get the prices that are displayed on the screen. The only way to effectively do this is for you to have the program that you are running constantly check the pricing feed for these discrepancies—which happen more than you realize. You must take advantage of them each time.

The cryptocurrency arbitrage market is a valuable one. You make a profit doing it while keeping the market balanced. It is an almost risk-free endeavor and it is one that is easy to program. The more currency pairs that you do this for, the more opportunities you will get in the mismatched pricing. You should have your trading program do this for all the counters that you can—even the ones that you are not totally familiar with—because this trade is not about the understanding of the nature of the currency, but rather takes advantage of the mispricings of any market.

There are a total of 11 cryptocurrencies (BTC plus 10) plus the USD that you should set your system up for. Then you keep an eye out for anything that starts to trade heavily and have an impact on the market It is still early days in the cryptocurrency world, and you can stand to make gains if you just do nothing but set your program to look for pricing mismatches and arbitrage opportunities.

OBV Mismatch

Trading on an OBV mismatch is a trade of the second order in the sense that you are no longer just looking at the price of things to determine a trade. You are actually looking at the effect of the activity around it that is measured by the OBV. When you see there is a mismatch between two counters, then you look at the price that they are trading at and see if the price of the underlying asset is converging or diverging. If it is converging, then you are setting yourself up for a sell order. If it's diverging, then you are setting yourself up for a buy order. This is one of the indicators that you can use after you get a handle on how to trade multiple currencies at once.

These strategies will help you get started with the cryptocurrency trading that should make it fun and interesting. Trading cryptocurrencies is about speed, agility and having the nerve to watch the fluctuations in the price and take advantage of it whenever possible. It is a highly opportunistic market and that is the mindset that you should bring to the table. This is not the best market for you to be relaxed with. If you bring your high-energy game to the table, you will be rewarded.

Mindset

Your mindset is the critical element that controls whether you are successful in this endeavor or you end up losing. The folks I see lose money in the market are typically from two categories. The first are the ones who do not bother to read up, and only slowly start to understand what the market is and what it is all about. You are obviously not in that category; after all, you are reading this to enhance your knowledge. So that's a brilliant start. But there is a long way to go.

The second category of folks are the ones who are not able to stomach the constant gyrations and fluctuations of the market. This is indeed not a slow market and things can turn around in a moment.

Then there are the three kinds of people you find when the market switches directions on them.

The first kind panic and liquidate. The second kind freeze and watch their holdings erode. And then there is the third kind that sees the change in direction as a potential for profit and as an opportunity, and the turn right around with it and reverse their exposure. You need to be quick on your feet and not fall in love with a position.

Psychologically, my motivation coach tells us that this is about not wanting to admit that you are wrong on a trade. I don't have that problem. I will be the first one to admit that I made a wrong call and switch positions so that I can be making money while the other guy is churning in the pit of his stomach waiting for the market to rebound. By the time it has fallen and rebounds and his position is even, I've made money twice.

The point I am trying to make is that you have to have a cool head while trading. And because you are human, the best thing to do to maintain this state, is to make sure you do programmed trading.

When you do programmed trading there are different strategies that can be programmed into the system. Your first exposure to the market will work if you can just get the four strategies mentioned in this chapter up and running.

Be sure to understand that there are some trades in which you will have to cut your losses. Further, when you cut your losses, you have to effectively and efficiently move on to the next one without worrying too much over spilt milk.

Traders, regardless of the instrument and underlying

asset, are a different breed all together. I have been in it for more than twenty years, trading currencies and watching the market for twenty out of twenty-four hours, five days a week. There is almost nothing that can happen in the market that scares me while I am in this position. That's due to the following factors:

1. My mindset during a trade and before the trading day begins is very simple. I am not emotionally connected to the trade. I do not celebrate when the trade is good, and I do not moan when it goes south.

2. I do not see the trade as money—I see it as a task and a response. It is a negotiation between the market and me. When the market swerves left, I move to counter. When it swerves right ,I move to counter again. And as long as I am in the market, I am constantly doing that without spending one moment on the thought that I am at the tip of however much money stands behind each call.

3. Each trade is cold. I work the numbers and I know the answer. It is the key to any trade. When you want to get into this market you have to know everything that is going on and you have to understand the tools that are at your disposal. That is the purpose of these books that you read. It is to get you to understand the tool that you have in the system and in the

market, as well as in your mind. They are all tools, and you have them at your disposal to be able to go out and execute a trade. I do take losses from one trade to the next but I make sure that at the end of the day the net position I close with is always positive—that's a mindset. If it isn't, it doesn't matter because I have tomorrow. I just look at the mistakes I've made and make sure to learn enough not to repeat them.

Chapter 8

Storing Your Coins

Here is my secret on how to store your coins. The first thing you do is make sure you have an spreadsheet that details each coin that you own. Make sure that all the codes are stored in your wallet, cold-stored and air-gapped. I will show you in this chapter how to air-gap them effectively.

Effective Air gapping

You need to have two computers and two storage systems. The first computer is the computer that you use to get online and trade. That computer does not hold all your coins under any circumstances. Never keep your unexposed coins on this machine.

Set up a computer system that is not linked to the internet in any way. It should not even have a Wi-Fi system or ability. This computer is not supposed to connect to other computers, so make sure you set it up in the operating system that the computer is not meant to network to anything—including printers and hotspots. On that computer, you can keep your wallet with the bulk of coins and make sure you have a duplicate copy in a thumb drive. Keep the thumb drive that holds the wallet.dat file in a safe. When you withdraw coins from your brokerage and store them for safekeeping, place the coins in the wallet.dat file and transfer it to the thumb drive, then send it to the computer. From the computer that is off-line, transfer it to the thumb drive that you place in your safe, as a copy.

The reason I have extra layers of protection and have them all virus-protected, is so there is no chance of being traced or tracked. The wallet that holds the bulk of my coins is kept off-line so that there is no way to see where it is located. Once I move the DAT file off-site, the computer will no longer pose a threat in case of hacking.

Proper Apportioning

The best form of security is to always keep your computers up to date with the necessary patches and the best (paid-for) anti-virus programs. You should also take precautions to keep your DAT files scrambled. Just remember to unscramble them when you place the wallet back in the computer and when you go back online. If you keep your coins properly separated,then you will have no risk of having them stolen. There are a number of people who have had their computers hacked with the coins in them, and while their computer shut down and was no longer accessible due to ransomware—so even the thieves did not get the coin—the coins were still lost for good. Remember, there is no way of recovering lost coins. You need to be responsible for them at all times.

You should also print out the coins and keep that printout in your safe. In the event that you do have a total burn-up of your hard drive or you have lost your thumb drive, at least you will be able to repopulate your wallet by typing in the coin's codes. Make sure to encrypt the file that has the coins.

Off-site storage

I never give anyone, any company or brokerage all my coins for safe keeping. I think that is the dumbest thing in the world to do. There have been hundreds of millions of dollars' worth of coins already stolen from seemingly trustworthy exchanges and there will continue to be more. It is the nature of cryptocurrency. It can't be traced and there will always be the temptation to sweep everyone's coins. It is the ideal scenario, don't you think?

If you have one hundred thousand users with 20,000 dollars' worth in each account and you supposedly hack all of them, that is 2 billion dollars' worth of coin (of course this is only for illustration purposes). No one is going to chase up the loss of just 20,000 dollars. But in all, the bandits get away with billions. So you see, it is human nature to take it if they know for a fact that they will not get caught. There is so much upside for very little risk. Stealing your bitcoins is the best return on risk there is in any market. I know this very well, and for this reason I will never place my coin in anyone's wallet except my own.

Only place what you must in a brokerage account, and if you can get an account where you place cash as a deposit and you are allowed to trade on that, then all the better. Every time your profit comes back in coin, transfer that coin to your own off-line wallet and store that. It is a good way to keep your assets safe and to be able to slowly convert your holdings as time goes by and the security of the coin of your choice increases.

But remember never to exceed your own rule of how much coin you should keep at all times. Use proper portfolio habits and practices, and never convert all your assets to coin—no matter how tempting that prospect might be.

Chapter 9

Why trade Bitcoin?

This book, while talking about cryptocurrencies in general, carries with it the implicit preference of using BTC as the trunk of all transactions. The lion's share of the portfolio of cryptocurrencies should be diverted to BTC instruments and traded—as said before—on a swing or day trade basis. In fact, the swing trades are not the priority. If you can, stay with day trades in BTC. But if you must do swing trades, then do not go beyond BTC. Under no circumstances are the altcoins ready for the kinds of risk that you would want to face.

The thing about risk is that you need to manage it and you need to realize that when managed well, it won't bite you, except for small losses here and there that you can cut and switch. However, the problem with holding swing positions is that you may be away when the market news hits that could turn your position against you.

The reason you want to keep BTC as part of your portfolio is that BTC is the trendsetter and it is the de facto core currency. It also has the highest possibility of becoming a mainstream currency. And if it does, the capital appreciation on that will be stratospheric. Think about that for a second. It only takes one country to take BTC as its sovereign currency and by virtue of that, the currency would have to be traded across the board. That alone would make the demand for that currency skyrocket as the number of ownerships will increase a thousand-fold as new owners will have to come in and the circulation will exceed even the wildest imagination of BTC enthusiasts now.

The underlying trend of BTC is fueled by that very knowledge. Those on the inside of the BTC game understand that this is indeed a very real possibility and that would make all the early adopters extremely wealthy. It will be one of the most significant events in modern history and in all of the world's financial history. And it could happen at any time now.

In the midst of all this, of course, there's a lot of irrational exuberance that causes the market to sway and gyrate. This is to be expected. If you have traded stocks before, or even options and currencies, you already have existing skill sets that can be transferred over. So you could potentially stand to make a significant profit in this market as long as you keep it strictly as a day trading or swing trading opportunity.

If you are, however, looking for a large upside and are looking to make the huge leap that Bitcoin had—from 8 cents to 17,000 dollars, then there are a couple of other newer currencies that you could try within the top ten that we talked about earlier. But do not let your glazed-over eyes obscure the real truth—that the risks are very real and it could go from thousands to zero in an instant without making any stops along the way. If you are leveraging yourself, that could be a real problem, and we will see why in the next chapter.

Chapter 10

Leverage

When you get to the point that you know what you are doing, then the next level is to leverage your positions. Remember in the earlier chapter on trading strategies we saw a simple day trading program that returned almost $2000 for the day. Giving this some deeper thought . . . that $2000 was a return based on the full purchase value on one lot of a particular currency. If you buy Bitcoin at $10,000 per unit and your profit is $2000 for the day, that's $10,000 per week, or $520,000 per year making it a 5100% return. And that's not even compounded.

If you were to increase your trade volume as you make more profit, that would have a compounding effect. But that's not as much as you will make if you leverage your position. But before you do, make sure you have the ability to cover the shortfall and keep trading otherwise you could be in a position of a margin call.

To trade on leverage is to trade on margin in BTC. The more reputable brokers allow up to three times leverage. That means if you want to realize an asset that is $12,000, all you need to put up is $4,000. So think of it in terms of the example earlier. If you put in $10,000 and got back $2,000 (actually it was

$1,820—I'm rounding up to make the calculation easier to follow) that is an annualized 5100% return. Now having said that, what if you only put in $3,333 (⅓ of 10,000) and leveraged the rest? That would mean your return would be 14,000%. The returns start to look unreal.

And yes, they are a little on the positive side and many of the realities that come into play will not show you a 14,000% return. But what does happen is after you factor in your bad days, your lost trades and your off days (days when you are on vacation or don't trade) you will realistically still experience positive return if you have the right moves and the right program running your trades.

Conclusion

That brings us to the conclusion of the first book in this series. What we have attempted to do is talk about the trading aspect of Bitcoin and the altcoins, and show you the merits of the cryptocurrency world and the development of this asset.

The cryptocurrency world has experienced a surge that has taken the pioneer of cryptocurrency, BTC, all across the globe and the imagination of enthusiasts. That widespread knowledge and acceptance makes BTC a lucrative trading instrument.

For the first time since the development of the study of economics and the advancement of international trade, we have a currency that transcends the limitations that fiat currencies have inherently built into them. From the fact that they have to be printed and that political motivations come into play when thinking about the amount to print, regardless of the -term consequence of diluting the world of fiats.

There is no danger of manipulation in cryptocurrencies as the currency cannot be made; it exists in mathematical quantities and cannot be created beyond that limit. It can, however, be accumulated and destroyed, and once destroyed it cannot be replaced. The risk of that happening now would be unfathomable. At the time of writing, the current market cap of BTC is in the region of about 200 billion dollars. If someone wanted to mop it all up, then the final price that they'd pay for it may be well over that, since the market price will appreciate as the purchases are executed. It makes no sense for anyone to spend 200 billion dollars just to wipe it all off. Also, look at it this way—if someone does do that, then your investment is covered because they would have to buy it from you as well, and it would be ostensibly higher than the market value.

But this is not investment advice. This is merely me sharing with you how I make money on BTC and what I would do if I had to start all over again. Compared to all the other currencies that I have traded, BTC and altcoins are one of the easier ones to make money with simple moves because they are so active and the market for them is rather rapid. There is no significant human element (as in physical criers and traders in a pit of the stock exchange) and the trades are fast and automated. If you could get the perfect program to do your trades, then you could conceivably put it on autopilot and just keep riding the market sentiment and waves.

Thank You!

Before you go, we would like to thank you for purchasing a copy of our book. Out of the dozens of books you could have picked over ours, you decided to go with this one and for that we are very grateful.

We hope you enjoyed reading it as much as we enjoyed writing it! We hope you found it very informative.

We would like to ask you for a small favor. Could you please take a moment to leave a review for this book on Amazon?

Your feedback will help us continue to write more books and release new content in the future!

Don't Forget to Download Your Bonus:

Bitcoin Profit Secrets

https://dibblypublishing.com/bitcoin-profit-secrets

CPSIA information can be obtained
at www.ICGtesting.com
Printed in the USA
LVHW111500190121
676890LV00031B/445